SPOTLIGHT ON NATURE

# BIGHORN SHEEP

NELL MUSOLF

CREATIVE EDUCATION · CREATIVE PAPERBACKS

Published by Creative Education and Creative Paperbacks
P.O. Box 227, Mankato, Minnesota 56002
Creative Education and Creative Paperbacks are imprints of The Creative Company
www.thecreativecompany.us

Design and production by Blue Design, Inc.
Art direction by Tom Morgan
Edited by Grace Beltowski

Photographs by Alamy Stock Photo/James Hager, 26; Corbis/23; Deposit Photos/jill@ghostbear.org, 8, 10, 11, 14, 16, 20, 22; Dreamstime/Anna Richard, 15, Sherry Young, 6; flickr/Biodiversity Heritage Library, 21; iStock/Petr Kahanek, 12, raincoast, 28; Pexels/Aaron J Hill, 16, 24, Britt Pigat, 22, bryan pascual, 18, Ricardo CL, 29 (top); Shutterstock/James Marvin Phelps, 9, 17, 29 (bottom), Montipaiton, 21, Tom Reichner, 10; Unsplash/Jon Sailer, 27, Paxson Woelber, 14, Pete Nuij, 4–5; Wikimedia Commons/Cary Bass-Deschenes, cover, 1, Rennett Stowe, 29 (middle)



Library of Congress Cataloging-in-Publication Data
Names: Musolf, Nell, author.
Title: Bighorn sheep / Nell Musolf.
Description: Mankato, Minnesota : Creative Education /Creative Paperbacks, [2026] | Series: Spotlight on nature | Includes bibliographical references and index. | Audience: Ages 10-13 | Audience: Grades 4-6 | Summary: "Spring into the bighorn sheep's world with this nature title for middle-grade wildlife lovers. Informational text pairs with a narrative about a single bighorn family to spotlight the wild animal's life cycle, supported by infographics and a timeline of developmental milestones"— Provided by publisher.
Identifiers: LCCN 2024048391 (print) | LCCN 2024048392 (ebook) | ISBN 9798889896180 (library binding) | ISBN 9781682777848 (paperback) | ISBN 9798889896982 (ebook)
Subjects: LCSH: Bighorn sheep—Juvenile literature.
Classification: LCC QL737.U53 M888 2026 (print) | LCC QL737.U53 (ebook) | DDC 599.649—dc23/eng/20241218
LC record available at https://lccn.loc.gov/2024048391
LC ebook record available at https://lccn.loc.gov/2024048392

Printed in India

# CONTENTS

MEET THE FAMILY

# BIGHORN SHEEP of Rocky Mountain National Park

It's a mild spring day in Rocky Mountain National Park. Located in northern Colorado, the park contains a portion of the Rocky Mountains. This famous mountain range stretches from northern Alberta and British Columbia in Canada down through the western United States to southern New Mexico. Bright purple horsemint grows in clusters where the sun shines. Herds of elk can be seen grazing in the valleys. Bears, moose, and deer are also commonly spotted in the jagged peaks and forested mountains of the park, but another resident is harder to find: the bighorn sheep.

On this special day, a **ewe** has left behind the herd. She chooses an isolated, sheltered spot where she has a clear view of her surroundings and any approaching predators. She must make sure this space is safe. After six months of preparation, the time has come for her to give birth.

CLOSE-UP

## Sensitive sniffers

A bighorn's sense of smell is quite strong, and it can identify different predators by their scents. All bighorns use their sense of smell to find water and food. Ewes use smell to locate their lambs.

CHAPTER ONE

# LIFE BEGINS

Bighorn sheep are a species of wild sheep native to North America. Approximately 80,000 bighorns can be found throughout Canada, the United States, and Mexico. There are three subspecies of bighorns: Rocky Mountain, Sierra Nevada, and desert bighorn. The Rocky Mountain bighorn is the largest in size and number, found throughout the Rocky Mountains. Sierra Nevada bighorns live in the Sierra Nevada mountains of California. Desert bighorns live in Mexico and the American Southwest. They are commonly found in the Mojave and Sonoran deserts.

The most distinctive feature of the bighorn is its large, curved horns. These horns are primarily used for fighting. They can weigh up to 30 pounds (13.6 kilograms). As **rams** age, their horns can grow as long as 3 feet (0.9 meter) with a 1-foot (0.3 m) circumference at the base. Ewes also have horns, but they are shorter with only a slight bend in them. The horns on a desert bighorn have tight curls. Sierra Nevada bighorns have a less pronounced curl in their horns.

ROCKY MOUNTAIN BIGHORN MILESTONES

**DAY 1**

- Born
- Wooly, white coat
- Small horns
- Can walk within hours
- Weight: 9 pounds (4.1 kg)

Bighorn sheep have compact, muscular bodies. Their fur can vary in color from almost white to dark chocolate with white trim on their muzzles, rumps, and bellies. The average male bighorn weighs between 160 and 250 pounds (72.6–113 kg). Some rams may get as large as 350 pounds (159 kg). Females are slightly smaller with an average weight between 125 and 150 pounds (56.7–68 kg). Desert bighorns are the smallest species and have adapted to the heat of the desert with less dense coats.

Bighorns spend much of their time on narrow mountain ledges that can be as small as 2 inches (5.1 centimeters) in width. They sleep on steep, rocky slopes. The ledges and slopes provide protection from predators that can't navigate the terrain like the bighorn. Bighorn predators include coyotes, golden eagles, mountain lions, bears, and Canada lynx.

CLOSE-UP

### Wide eyes

With eyes placed on the sides of the head, bighorn sheep have a wide arc of vision. They can see predators without turning. Bighorns can spot a predator from up to 1 mile (1.6 kilometers) away.

FEATURED FAMILY

## Welcome to the World

**In the Rocky Mountains, the ewe is ready to become a mother. Ten days earlier, her teats began to harden as they filled with colostrum. Her body was telling her it was time to give birth, so to prepare, she is perched on a cliff, safe from predators and alone until her lamb is born. The lamb arrives, white and covered with wool. He is born with small horns. He will be able to walk by the end of the day, but his mother will keep him hidden for more than a week for his protection.**

As **ruminant** herbivores, bighorns live on a diet of grasses, **herbaceous** plants, and shrubs. Clover is a favorite food, as are willow and sage. Bighorns in desert regions may also eat holly and cactus. Bighorns are taught to be selective eaters from a young age since food in the mountains can be scarce. They learn to dine on new plant growth, which is typically the most **nutritious**.

**1 WEEK**

- Lives on mother's milk
- Joins mother's ewe herd
- Starts climbing up steep surfaces

**3 MONTHS**

- Drinks less milk
- Grazes at mother's side

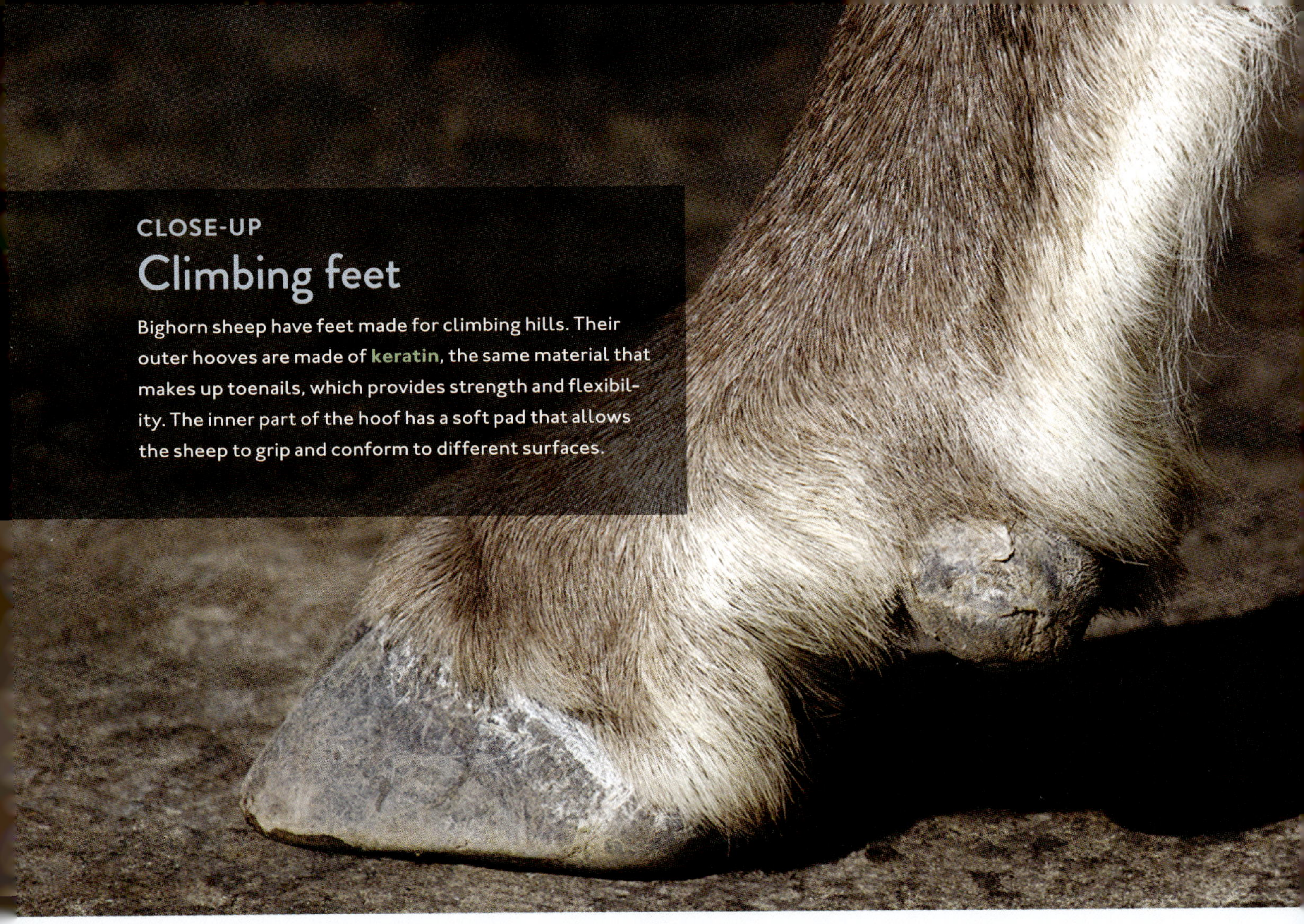

CLOSE-UP

## Climbing feet

Bighorn sheep have feet made for climbing hills. Their outer hooves are made of **keratin**, the same material that makes up toenails, which provides strength and flexibility. The inner part of the hoof has a soft pad that allows the sheep to grip and conform to different surfaces.

FEATURED FAMILY

## First Meal

The newborn lamb drinks his mother's milk. The milk provides all the nutrients necessary for the first few months of life. Nursing also creates a strong bond between the ewe and her lamb. The young lamb will be **weaned** between three and six months, but within a few weeks of birth, he will already begin eating the grasses, flowers, and plants that will make up his diet for the rest of his life. His mother will be at his side, guiding him.

An adult **BIGHORN SHEEP** can have horns that weigh up to 30 pounds (13.6 kg).

### 6 MONTHS

- Fully weaned from mother
- Milk teeth are present
- Horns are still not developed
- Weight: 75 pounds (34 kg)

CLOSE-UP

## Four stomachs

Bighorn sheep have a four-part stomach. This allows them to swallow big portions of food quickly. Then, they go back to their ledge or cliff dwelling, where they can spit up the food and rechew it in safety.

CHAPTER TWO

# EARLY ADVENTURES

The most important survival skill a bighorn sheep learns is its climbing ability. Lambs can run and climb within a few days of birth and have rubbery hooves that help them move up mountainsides. As they grow, they improve their climbing skills by competing to see who can run the farthest and fastest up steep slabs of rock. Their agility in climbing up mountainsides and their adaptable foot pads help the sheep escape predators.

Lambs stays close to their mothers in ewe groups, where they learn how to find food. Sierra Nevada and desert bighorns eat buckwheat and a variety of grasses that grow in snow-free areas. Rocky Mountain bighorns dine on grasses, sedges, and clover during warm months. During winter months, their diets change to willow, holly, cactus, and sage. Living in the mountains can be dangerous during the winter. Heavy snows can produce avalanches and rockfalls on the cliffs where bighorns live. Some herds avoid

**1 YEAR**

- Permanent teeth start to emerge in lower jaw
- Horns begin growing
- First ring forms on horns
- Body length: 32 inches (81.3 cm)

CLOSE-UP

## Excellent ears

Bighorn sheep have excellent hearing. They can direct their ears toward a noise. Sound waves arrive at slightly different times and volumes to each ear. Bighorn sheep are alarmed by high-pitched noises and loud sounds like barking and yelling.

FEATURED FAMILY

## Eat Up!

The three-month-old lamb is ready for lunch. His mother has been nursing him since birth, and he likes the convenience of being able to dine on demand. But the time is quickly coming when his mother will no longer be willing to nurse her lamb. The mother sheep knows her body will soon stop producing as much milk so she can start storing fat to survive the long winter. But first, she wants to fatten up her lamb as much as possible. Lambs who carry more weight will survive winter a little more easily.

such dangers by moving to lower elevations during the winter where there is less snow and more plants to eat.

Avoiding predators and finding food aren't the only challenges bighorns face. Bighorn sheep are highly susceptible to **respiratory** diseases. These diseases are caused by **pathogens** carried by domestic sheep. Domestic sheep were introduced in the western part of the United States during the 1800s and brought diseases that have killed many bighorns. Conservationists learned the best remedy is to avoid interactions between domestic and bighorn sheep. That isn't always possible in the wild, and the risk of losing bighorns to respiratory diseases remains a constant threat.

**2 YEARS**

- Males and females reach sexual maturity
- Females mate for the first time
- Horns continue growing and begin to curl

CLOSE-UP

## Big horns

The horns of bighorn sheep are not just decorative. Along with age and body size, horn size determines which rams are herd leaders. The bigger the horns, the more control a ram has over the herd.

FEATURED FAMILY

## Give It a Try

Yuck! What is that? The young lamb spits out a mouthful of grass that looked good but tasted dry and awful. His mother warned him about eating something dried out. He should have known to stick with the plants his mother showed him. The lamb moves on to a patch of clover. It is sweet, damp, and delicious. After he finishes with the clover, maybe he can find some sedge to chew on. Mother really did know best.

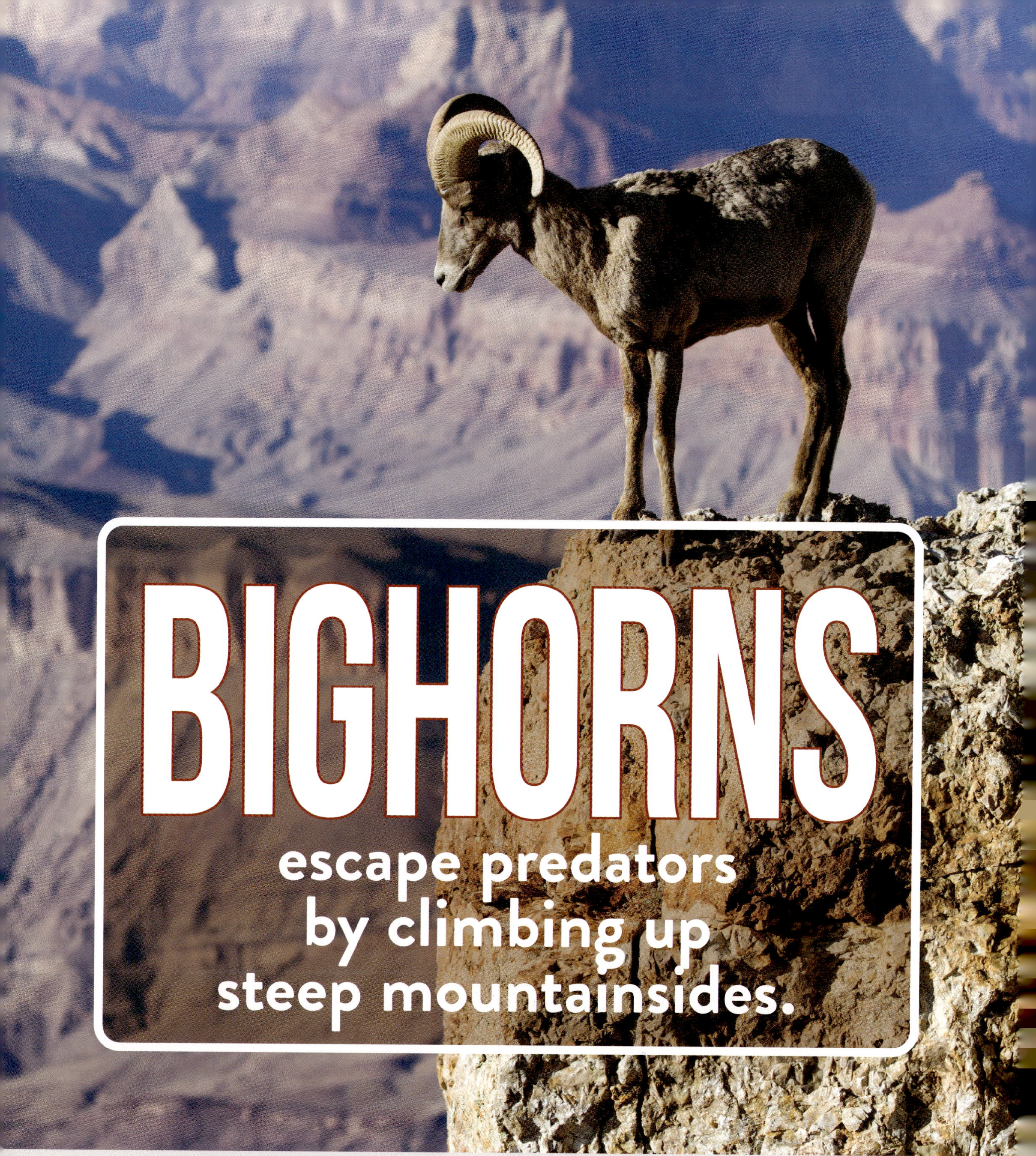

# BIGHORNS

## escape predators by climbing up steep mountainsides.

④ YEARS

- Males have joined an established ram herd
- Males may compete in headbutting but lose
- Dark rings begin to appear on rams' horns
- Female horn growth slows

CLOSE-UP

## Fur

Bighorn fur varies from off-white to dark brown. These colors help them blend in with their environment. Bighorns have heavy, double-layered winter coats that keep them warm in the colder months. They shed their winter coats in the summer.

CHAPTER THREE

# LIFE LESSONS

When male bighorns are between two and four years old, they leave their mother and join a herd made up of other rams. Eventually all males must join a new herd, and it's up to the ram to find one. This can be a difficult time of wandering alone until a group of rams lets him in.

Ewes and rams are able to begin breeding around two years of age, but rams must typically wait until they are older and their bodies are fully developed. Their horns grow with age and determine their place in the herd. Younger rams with smaller horns are less powerful, so they are rarely able to mate before seven years. The mating season is also known as the "rut" and occurs between August and January, depending on the species. During the rut, rams use their horns to battle each other in headbutting competitions. The rams become very vocal and snort and grunt. These competitions can be violent and often result in broken horns. Rarely the competitions end in death. The winner of the competition becomes more powerful than his opponent and has more control over the herd.

**6 YEARS**

- Runs at 30 miles (48.3 km) per hour on level ground
- Climbs uphill at 15 miles (24.1 km) per hour
- Height: 40 inches (102 cm)

Following headbutting competitions, the strongest ram usually has several ewes to mate with. The ram shows his interest by rubbing his scent on foliage and rocks. He also follows the ewes closely. His actions mark his territory and show the ewes that he is ready to mate. Lambs are born the following spring. Ewes generally give birth to one lamb at a time.

During the rest of the year, rams remain in their male herds and don't mix with female groups or young sheep. Female bighorns remain with their mother's herd for their entire lives. Their herds consist of ewes and rams under two years of age, and they are led by an older, mature ewe. Both male and female herds usually have between 8 and 10 bighorns. Herds can have as many as 100 sheep, but that is rare. Bighorn sheep have an average life expectancy of 8 to 12 years.

FEATURED FAMILY

## This Is How It's Done

**The ram is now three years old. No longer living with his mother, he has joined a herd of other rams. It's grazing time. He drifts away from the herd as he looks for his favorite grasses. Out of the corner of his eye, he sees something moving. Lifting his head, he sniffs the air, identifying the presence of another animal: a coyote. One last swallow and the ram leaps to a nearby ledge where he is safe and can finish his meal in peace.**

CLOSE-UP

## Complicated skulls

Unlike in humans, the plates in the skulls of bighorn sheep have jagged edges that do not fuse together completely. The plates are a little looser and allow the rams to bang into each other without causing major injury.

### 7 YEARS

- Horns are fully curled
- Hairless knob forms on back of rams' heads
- Males win headbutting competitions
- Males mate for the first time
- Females continue birthing one lamb each year

CLOSE-UP

## Cough, cough

Most of the time, bighorns are quiet creatures who rely on body posture and scent to communicate with each other. When alarmed, a bighorn may make a coughing noise to alert other bighorns. They also occasionally bleat.

FEATURED FAMILY

# Practice Makes Perfect

The grass tasted good when the ram first pulled it out of the ground. He was hungry and ate a lot, but he had to eat quickly in case there were any dangerous animals nearby. The grass went into the first part of the bighorn's four-part stomach. Later, alone and safe on a rocky slope, the ram **regurgitates** the grass. He chews it again more leisurely, now having the time to enjoy his meal completely.

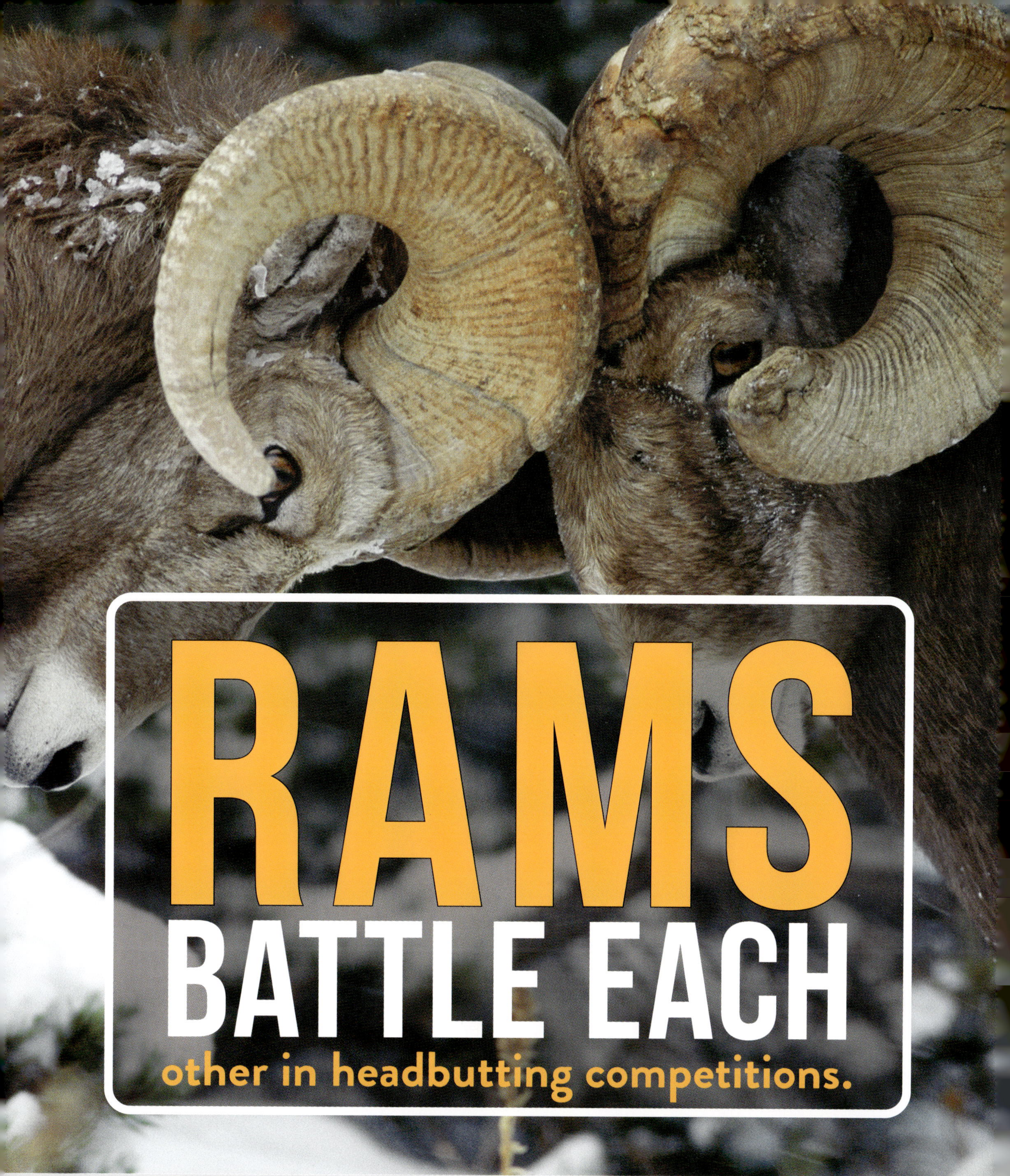

# RAMS BATTLE EACH other in headbutting competitions.

(12) YEARS

- Horn length: up to 3 feet (0.9 m)
- Horn weight: up to 30 pounds (13.6 kg)
- End of life

CHAPTER FOUR

# HELPING BIGHORNS SURVIVE

Humans are a major threat to bighorn sheep. For many hunters, a pair of bighorn horns is a highly sought trophy not only because of their unique look, but because the bighorn is such a difficult prey to catch. Prior to the American West being settled, the number of bighorn sheep in the United States was estimated to be between 150,000 and 200,000. As more people arrived in the area, the bighorn suffered declines and the threat of **localized extinction**. By 1860, the North American bighorn population had dropped to between 15,000 and 18,000.

To preserve bighorn populations, many states enforced limited hunting seasons. In Montana, the first conservation law was passed in 1869. Bighorn sheep could only be hunted between August and January of each year. Even with such efforts, an entire subspecies of bighorn sheep, the Audubon bighorn, went extinct by 1925. The 1930s saw the passing of the Federal Aid in Wildlife Restoration Act. This act set a tax on hunting equipment and guns. The tax was used for the restoration of wildlife resources and habitats. Despite the Wildlife Restoration Act, the population of Sierra Nevada bighorns was down to two herds in the 1970s. Currently, this subspecies of bighorns is considered endangered in the United States.

Thanks to **translocations**, bighorns have been moved to safer locations and their population has slowly grown to approximately 65,000 in North America. There are ways people can ensure that bighorns are around for future generations. One of the best ways is to ask representatives and senators who represent the western states to make wise land use decisions. A more personal way to help is by avoiding trails during lambing season. Having a safe space to give birth is crucial to the future of bighorns. Avoiding camping near water sources in the wild is another way to help bighorns and other wildlife. Cleaning up after hikes and camping trips shows respect for bighorn sheep and all creatures who call the great outdoors home.

FAMILY ALBUM

# SNAPSHOTS

Wyoming is home to the largest population of **Rocky Mountain bighorns**—approximately 6,000—in the lower 48 states.

Rocky Mountain bighorns are often used as symbols for Rocky Mountain National Park in Colorado.

**Sierra Nevada bighorns** have adapted to dry desert climates and can lose 30 percent of the fluid in their bodies before becoming dehydrated.

Although Sierra Nevada bighorns have the smallest horns of all bighorn species, their horns are wider and stick out more.

**Desert bighorns** have shorter lifespans than other bighorn species and typically live between six and eight years.

Desert bighorns are the only bighorn sheep found in Mexico.

Bighorn sheep can jump distances of 20 feet (6.1 m). Their excellent climbing skills are second only to mountain goats.

Bighorns prefer high altitudes and rugged weather conditions. They live at elevations of more than 10,000 feet (3,048 m) above sea level.

You can tell the age of a bighorn by the size of its horns and the number of rings on it. One ring equals one year.

# WORDS to Know

**colostrum** the first secretion from mammary glands after giving birth

**ewe** a female sheep

**herbaceous** relating to herbs

**keratin** a fibrous protein that makes up hair, feathers, horns, claws, and hooves

**localized extinction** the disappearance of a race or species from a specific geographic region

**nutritious** nourishing or effective as food

**pathogen** a tiny organism that causes disease

**ram** a male sheep

**regurgitate** bring swallowed food up again through the mouth

**respiratory** relating to breathing or organs of respiration

**ruminant** a mammal that chews its partially digested food after spitting it up

**translocation** the movement of something from one place to another

**wean** accustom an infant to food other than its mother's milk

# LEARN MORE

## Books

Gagne, Tammy. *Bighorn Sheep*. Lake Elmo, Minn.: Focus Readers, 2017.

Leaf, Christina. *Rocky Mountain National Park*. Minneapolis: Bellwether Media, 2023.

Riggs, Kate. *Bighorn Sheep*. Mankato, Minn.: Creative Education and Creative Paperbacks, 2023.

## Websites

"Bighorn Sheep." Montanakids.com. https://montanakids.com/plants_and_animals/animals/bighorn_sheep.htm

"Bighorn Sheep." National Geographic. https://www.nationalgeographic.com/animals/mammals/facts/bighorn-sheep

"Bighorn Sheep." Washington NatureMapping Program. http://naturemappingfoundation.org/natmap/facts/bighorn_sheep_k6.html

## Documentaries

Green, Frank, dir. *Counting Sheep: Restoring the Sierra Nevada Bighorn*. Stonington, Conn.: Green Planet Films, 2004.

*Running with the Herd: A NATURE Short Film*. Alexandria, Va.: PBS Nature, 2018.

Town of Vail, dir. *Home for the Wild*. Vail, Colo.: Capture the Action Productions, 2022.

Note: Every effort has been made to ensure that any websites listed above were active at the time of publication. However, because of the nature of the Internet, it is impossible to guarantee that these sites will remain active indefinitely or that their contents will not be altered.

# Visit

### BUFFALO ZOO

*Bighorn sheep have their own exhibit at this New York zoo.*

300 Parkside Avenue
Buffalo, NY 14214

### HEMKER PARK & ZOO

*See bighorn sheep in the Midwest while on a tram ride that goes all over the zoo.*

26715 County Road 39
Freeport, MN 56331

### ROCKY MOUNTAIN NATIONAL PARK

*Observe bighorn sheep in their natural habitat at Horseshoe Park. Late spring and early summer are the best times to spot bighorns.*

U.S. Highway 36 West
Estes Park, CO 80517

### SAN DIEGO ZOO SAFARI PARK

*Head to the Condor Ridge exhibit to view desert bighorn sheep.*

15500 San Pasqual Valley Road
Escondido, CA 92027

# INDEX